Beneath Springhill:
The Maurice Ruddick Story

Beneath Springhill: The Maurice Ruddick Story

Beau Dixon

Lyrics and Music
by Rob Fortin and Susan Newman

Beneath Springhill: The Maurice Ruddick Story
first published 2021 by Scirocco Drama
An imprint of J. Gordon Shillingford Publishing Inc.

Scirocco Drama Editor: Glenda MacFarlane
Cover design by Doowah Design
Photo of Beau Dixon by Tim Leyes
Photo of Rob Fortin and Susan Newman by John Lewis
Production photos by Nicole Zylstra
Printed and bound in Canada on 100% post-consumer recycled paper.

We acknowledge the financial support of the Manitoba Arts Council
and The Canada Council for the Arts for our publishing program.

For production information, permission to use the
original music, or bookings, go to:
www.firebrandtheatre.com or www.beaudixon.com

Title: Beneath Springhill : the Maurice Ruddick story / Beau Dixon.
Names: Dixon, Beau, 1973- author.
Description: A play.
Identifiers: Canadiana 20210138823 | ISBN 9781927922767 (softcover)
Subjects: LCSH: Ruddick, Maurice, 1912-1988—Drama. | LCSH: Springhill Mine Disaster, Springhill, N.S., 1958—Drama.
Classification: LCC PS8607.I93 B46 2021 | DDC C812/.6—dc23

J. Gordon Shillingford Publishing
P.O. Box 86, RPO Corydon Avenue, Winnipeg, MB Canada R3M 3S3

To my family:

Mom, Dad, Lisa, Lance, Jade, Nathan, Ray, Kalista, Mercy, Soren, Kelli, Mark and Karen.

To the Ruddick family:

Maurice, Norma, Colleen, Sylvia, Valerie, Alder, Ellen, Dean, Francis, Revere, Leah, Jesse, Iris, Katrina and Maureen. Doug Jewkes, Currie Smith, Garnet Clarke, Herb Peperdine, Frank Hunter and Percy Rector.

To the people of Springhill: the ninety-nine miners who lived to tell their story — and to the seventy-five who lost their lives to the "Bump." This is for you...

The No. 2 Bump — October 23, 1958

Allen, Fidele
Aylward, Ralph
Backa, Andrew
Bobbie, Edward
Bourgeois, Bliss
Brine, Henry
Bryan, Percy
Burton, Charles
Canning, George
Cole, Cecil
Corkum, Clyde
Crowe, Hance
Embree, Harold
Embree, Harry
Fraser, Harold
Gerhardt, Joseph
Gillis, Angus
Goode, Kenneth
Halliday, Harry
Harrison, Cecil
Harrison, Chesley
Henwood, Harlan
Holloway, Isaac
Hunter, Hiram
Hunter, Wylie
Hyatt, Warren
Jackson, John
Jewkes, William
Leblanc, Abbey
Legere, Alfred
Livingstone, Gilbert
Macdonald, Arthur
Macdonald, Edward
Macdonald, Harold
Macfarlane, Roy
Mackenzie, Frank
Mackinnon, Edwin

Macleod, Charles
Macleod, Clarence
Macleod, Edward
Macleod, Frank
Macleod, Robert
Macleod, Varley
Mcnutt, Harold
Maddison, John
Marshall, Thomas
Miller, Bernard
Mooring, Carl
Nicholson, Fred
O'Brien, Harry
Perrin, Robert
Porter, Stirling
Raper, Harold
Rector, Percy
Reid, Joseph
Reid, Layton
Reid, Lester
Reynolds, Wesley
Rolfe, Ernest
Rose, Charles
Ross, Philip
Ross, Robert
Ross, St. Clair
Smith, William
Spence, Percy
Stevens, Eldon
Stevenson, William
Tabor, Hollis
Tabor, Monty
Tabor, Raymond
Teed, Henry
Turnbull, William
Welch, George
White, Albert
White, Carl

Beau Dixon

Beau Dixon is a multi-award-winning playwright, actor, composer, music director, and sound designer. As a playwright he has been commissioned by Arbor Theatre, Heritage Pavilion, 4th Line Theatre, Theatre New Brunswick, Nova Scotia Tourism, and Parks Canada / Canadian Citizenship and Immigration. *Beneath Springhill: The Maurice Ruddick Story* won a Dora Mavor Moore Award (Outstanding New Play / TYA Division).

As an actor, Dixon has received two Toronto Theatre Critics Awards, a Calgary Critics Award, and a Dora Mavor Moore Award. He is a KM Hunter Award finalist and was also inducted into Peterborough's Pathway of Fame for his leadership in the arts. Beau is artistic director and co-founder of Firebrand Theatre. He continues to write and perform, dividing his time between Peterborough and Toronto.

Rob Fortin and Susan Newman

Peterborough, Ontario-based husband and wife team Rob Fortin (lyrics) and Susan Newman (music) have spent decades together writing songs for plays, performing and music directing. Their work includes two full-length musicals: *Cavan Casanova* (book by Robert Winslow, 4th Line Theatre, 2003), and *Hungry: A Musical Hansel & Gretel* (book by Kate Story, Public Energy, 2008). In 2015 they were awarded, along with Beau Dixon, a Dora Mavor Moore Award (Outstanding New Play, TYA) for Firebrand Theatre's Beneath Springhill: The Maurice Ruddick Story, for which they composed the songs. They are currently developing a new musical, *Airborne*, with Beau Dixon and Michele Coleman. Susan is the director / arranger of the Convivio Chorus. Together they comprise two thirds of the jazz vocal trio Chester Babcock.

Acknowledgements

Thank you: Megan Murphy, Valerie (Babe) Ruddick, Barry and the MacDonald family, Rob Fortin, Susan Newman, Linda Kash (I'm forever grateful xo), Paul O'Sullivan, Caitlin Driscoll, Claire Levick, Rob, Tom and Charlie McInnis, Frank Morrison, Basil Chiasson, Sean Dixon, Doug Wyse, Philip Akin, Gabe Robinson, Esther Vincent, Alex Saul, Glenda Stirling, Diane Goodman, Ron Jenkins, Pamela Halstead, Mark Bellamy, Emma Brager, Samantha MacDonald and Lunchbox Theatre, Crystal Hoeg, Ashlie Corcoran and Arts Club Theatre, Kim Blackwell, Robert Winslow and 4th Line Theatre, Richard Rose and Tarragon Theatre, Erica Angus and Theatre Collingwood, Andrea Houssin, Djanet Sears, Andrew Moodie, Obsidian Theatre, Chris Oldfield, Marie Claude Valiquet, Craig Hall and Vertigo Theatre, Emma Slipp, Arkady Spivak and Talk Is Free Theatre, Mary Blackstone, Carol LaFayette-Boyd, Jaimi Williams, Kathryn Ricketts, Dan Fortin, Ontario Arts Council, Playwrights Guild of Canada, Glenda MacFarlane and Scirocco Drama.

Playwright's Notes

I grew up listening to music. I found great comfort lying on my parents' living room carpet and listening to their record collection. At night, my father would sit on the edge of my bed, strum his guitar and sing to me until I fell asleep. His voice always gave me comfort. It sounded like home.

In 2010, a friend approached me and told me about an African Canadian coal miner who was named Citizen of the Year for saving the lives of the last six surviving coal miners in the Springhill disaster. His name was Maurice Ruddick. He was a devoted husband to his wife Norma and a loving father of thirteen (their thirteenth baby, Maureen, was born two years after the Bump) living in Springhill, Nova Scotia. Maurice was a man of mixed race — like me. He also had a deep passion for singing — like me — and he had thirteen kids (my father had thirteen siblings). It would be fair to say that it was his singing that kept them alive while trapped two miles beneath the earth's surface for almost nine days. It would be fair to say that music has saved my life many times over. I knew I was destined to write a play about this heroic unsung hero.

The Springhill mining disaster, which occurred on October 23, 1958, was the most severe bump (earth tremor) in North American history. The Bump devastated the people of Springhill, Nova Scotia due to the casualties they suffered. It also devastated the town, as the coal industry had been its economic lifeblood. The disaster became famous for being the first major international event to appear on live television broadcasts (on the CBC).

As I interviewed friends and relatives of Maurice Ruddick, I learned a great deal about humanity, survival, and the strength of a community. Maurice's story of survival also encouraged me to ask some important questions: How do we deal with life and death decisions? Where does hope lie when we're faced with difficult choices? What does it mean to be human?...

Beneath Springhill has travelled throughout Canada. And now, it's in print for the world to enjoy. I hope Maurice's story inspires you as much as it inspired me.

Songwriters' Notes

Writing the songs for *Beneath Springhill,* we returned to our artistic roots. Our earliest collaborations as songwriters were for companies that practised popular theatre. Exploring stories of everyday people and their lives, we wrote songs set on picket lines and food bank lines, in a chainsaw factory, a post office, hospitals and a woollen mill, among others. We wrote songs about free trade, taxation, industrial pollution, and injured workers' rights. In almost every instance we had as raw material interviews and reminiscences of actual working folk. When Beau told us about his Springhill project, we thought that this was a story we could help tell, and we begged him for the chance to write Maurice Ruddick's songs.

As unregenerate old folkies, we had ready to hand the kinds of songs we thought would serve the story: field hollers, hymns and lullabies. When we wanted a song to sound like it was playing on Norma's radio, it couldn't hurt that we were both alive in 1958, listening to the radio.

As well, Rob's childhood, growing up in mining towns, gave him a personal route into the story. When it came time to imagine Maurice's predicament, Rob found it invaluable to be able to draw on his memories of stories told to him by his grandparents, uncles, his father, and brother of working underground. He can remember the sensation of the ground shaking on blast days, and the nightmare-inducing stories of accidents — particularly harrowing given the nuts and bolts of the industry: dynamite, the cage, "loose," poisonous or molten tailings — that are part of the fabric of mining towns. And their stories also gave him some of the language to tell the story, the detail necessary for the songs to feel true.

But the best tool we had to work with was the vivid picture of the coal miner Maurice Ruddick we were given, and that, of course, was all down to Beau Dixon. Sitting around the table listening to Beau breathe life into this play stands as one of our most exciting experiences as artists.

Maurice Ruddick was a genuine Everyman hero. It was a privilege to help tell his story. Thank you, Maurice and Norma, the baker's dozen, and, of course, Beau Dixon. We hope our songs have served you all well.

Susan Newman and Rob Fortin

Foreword

When supper was over and the dishes were washed, we went outside to play in the driveway and backyard. It was October 23, 1958, a clear, yet cool, fall night. It seemed every kid in the neighbourhood was outside playing hopscotch, marbles, hide 'n' seek, or taking turns riding around in an old cart with broken wheels. Back then we didn't have to be told to go out and play, we just did it. We were laughing, shouting, and teasing — when suddenly the ground shook and rumbled beneath our feet. Instinctively, we looked toward the mine that was within walking distance from our homes. Suddenly the town whistle blew, that same whistle that blew every night at nine o'clock to remind all kids that it was curfew time, time to be off the street and go home. But it wasn't nine o'clock yet; it was closer to 8:00 pm.

In fear we all ran to our homes, not knowing what was happening. In the living room, our mother, Norma Ruddick, was watching the end of an *I Love Lucy* show on a black-and-white TV. When we gathered around her, she was holding next to her bosom her twelfth child, then only two weeks old, and she had tears running down her face. She had a feeling there was trouble at the mine; she looked at all of us and said, "Your father is working the back shift at No. 2 mine."

The TV was left on, and — how ironic — the *Don Messer Show* appeared, and Marg Osborne and Charlie Chamberlain began singing: "Farther along you'll know about it, farther along you'll understand why, cheer up my brother…" This hymn seemed to bring comfort to our mother.

Amid the confusion and commotion, our neighbours brought news that something had indeed happened in the No. 2 mine.

It was soon confirmed by radio reports.

Forever etched in my mind are memories of me and my sisters going to the mine site where there were tents set up by the Red Cross and Salvation Army and others that provided comfort, food, and drinks, and hope and information on the rescue operations. We went every day, hoping and praying to bring home good news to our mother about our father.

In 2012, it was with great pleasure that I received a phone call from Beau Dixon (Babe) from Peterborough, Ontario. He informed me that he was writing a one-man play based on the mine disaster known as "The Bump" with emphasis on the role that my father, Maurice Ruddick — known as "the Singing Miner of Springhill" — played. Beau asked me to assist him with the background, and to tell him what I remembered in order to help him stage the play. Although at the time I had not met Beau in person, I trusted with confidence that his sincerity and interest would bring *Beneath Springhill* to life. During many calls with Beau, I was able to share with him and relive my story of The Bump and how it affected our lives.

It was so exciting to finally meet Beau in Springhill, NS, in February of 2013. I was honoured to introduce Beau to the audience that night before he performed his play to a full house. As Beau played each character, he allowed himself to enter the depths, two miles down into the pit of No. 2 mine. His body and soul were united and alive as we watched breathlessly. Beau captured in song those hymns as if he had written them himself, leading the other miners in hope and faith, crying out in prayer, and banging on the pipes — which was a daily ritual by the trapped miners, who were hoping to be heard by the rescuers. We, the audience, waited too: we cried and remembered the miners who died, and those who would live to see the light of day once again.

Valerie Ruddick
February 2021

Production History

Beneath Springhill: The Maurice Ruddick Story was originally produced by Firebrand Theatre in January 2014 as a touring show for young audiences. It received three Dora Mavor Moore Award nominations (TYA Division), winning for Best New Play and Individual Performance. The nomination for Best Director went to Linda Kash.

Performer: Beau Dixon
Lyrics and Music by Rob Fortin and Susan Newman
Directed and Developed by Linda Kash
Sound Design by Beau Dixon
Set Design by Gabe Robinson
Production and lighting design by Rob McInnis
Tour and stage management by Rob McInnis

Beneath Springhill: The Maurice Ruddick Story premiered on January 12, 2015, as a High Performance Rodeo production at Lunchbox Theatre, Calgary, Alberta, with the following cast:

Performer: Beau Dixon
Lyrics and Music by Rob Fortin and Susan Newman
Directed and Developed by Linda Kash
Sound Design by Beau Dixon
Set Design by Gabe Robinson
Production and lighting design by Rob McInnis
Additional Lighting Design by Anton De Groot
Stage Manager: Emma Brager

Beneath Springhill: The Maurice Ruddick Story premiered on July 19, 2016 at Thousand Islands Playhouse, Gananoque, Ontario

Performer: Beau Dixon
Lyrics and Music by Rob Fortin and Susan Newman
Directed by Linda Kash
Sound Design by Beau Dixon
Set Design by Gabe Robinson
Lighting design by Rob McInnis
Production and additional lighting design by Brian Frommer
Stage management by Marie Claude Valiquet

Beau Dixon as Maurice Ruddick. Lunchbox Theatre. Photo by Nicole Zylstra.

Time:

1958

Place:

Springhill, Nova Scotia

Cast

(All characters are to be performed by one actor.)

Family Members

Maurice Ruddick: 46 yrs.

Norma Ruddick: 33 yrs.

Valerie Ruddick: 10 yrs.

Journalist:

Jack McNeil: 40 yrs.

Miners

Doug Jewkes: 40 yrs.

Frank Hunter: 45 yrs.

Percy Rector: 55 yrs.

Garnet Clarke: 29 yrs.

Currie Smith: 36 yrs.

Herb "Pep" Pepperdine: 33 yrs.

Note: Although the play is the author's fictionalized account of the events that occurred in October of 1958, he has used the actual names of miners and others affected by the "Bump" in tribute to their humanity and resilience.

Scene One

October 31, 1958. Springhill, Nova Scotia.

Two miles below the No. 2 Cumberland Mine, a coal miner named MAURICE RUDDICK crawls out of the dark.

MAURICE: OH DEAR LORD, IS THERE ANYONE THERE?

ANYONE TO HEAR THIS MINER'S PRAYER

TRAPPED UNDERGROUND WHEN THE GREAT BUMP ROARED

IN THE COAL BLACK DARK

CAN YOU HEAR ME, LORD?

TRAPPED IN THE DARK TWO MILES UNDERGROUND

TWO MILES BENEATH THE GOOD FOLK OF SPRINGHILL TOWN

OH DEAR LORD, HEAR THIS MINER'S PRAYER

IS THERE ANYONE THERE?

IS THERE ANYONE THERE?

Please God. Help me. Say this isn't the end. Tell me I'll see my wife. My kids. My freedom....

Turns his head.

Who's there?...

There is the sound of an explosion and fallen debris. In the distance we hear sirens.

Lights up on a reporter, JACK MCNEIL, who is adjusting his microphone and ad libbing with a sound operator: "How are the levels? Can you hear me?... Standby... Three...Two..."

CBC theme music is heard.

JACK: This is Jack McNeil coming to you live on this cold day of October 23rd, 1958, in Springhill, Nova Scotia. News has just come in that an earth tremor has caused a "bump" in North America's largest coal mine, The Cumberland No. 2. It is now reported that one hundred and seventy-four men are trapped thousands of feet below the earth's surface. It is hard to say how many are dead and how many are alive. Even though this small community is used to catastrophic events — the mining disaster of 1938, the mining disaster of '56 — this is the largest disaster ever. As we keep you posted on the current events, we ask that you hope and pray for the families and friends who are waiting anxiously for the outcome of their husbands, their brothers, their fathers, their sons. This is Jack McNeil, CBC News, Springhill, Nova Scotia.

Scene Two

Springhill, Nova Scotia. 1958. Six months after the bump. MAURICE RUDDICK addresses an audience.

MAURICE: Good day to you. I'm Maurice. Maurice Ruddick. I'm forty-six years old, and I'm an African-Canadian living in Springhill, Nova Scotia with my wife and twelve kids. Some people call me "The Singing Miner"...

Clears his throat and begins to sing:

WAY WAY DOWN DIGGIN' DOWN IN THE DEEP

I'M A COAL-DIGGIN' DADDY DIGGIN' COAL FOR MY KEEP

FILLING BOX AFTER BOX

THAT'S HOW I EARN MY PAY

IT'S DOWN UNDERGROUND I SING MY BLUES AWAY

...Other folks call me a mulatto. Person of mixed race. *(Beat.)* Some people call me a nigger. *(Beat.)* But, I prefer you just call me Maurice. *(Beat.)*

Yup, ever since I could remember I wanted to be a musician. I would be that kid at church waiting for the choir hymns to be called out. I knew them all by heart. *(Plays a guitar lick.)* Yeah, I wanted to sing under the big bright lights. But, when you have a wife and twelve kids to feed...Well, sometimes you have to put those dreams aside.

So, I became a coal miner. Well, that's what you did in Springhill, you were a coal miner. That's what my father did and his father before him. Oh — I was born in the town of Joggins, just twenty-eight miles outside of Springhill. On my days off — if you didn't see me wrestling with one of my kids — you could find me hiding out in the backroom singing and writing songs on my trusty guitar. On some nights — if the night was just right — I'd tippy-toe to my little ones' bedroom and sing them to sleep.

He begins to strum on his guitar and sing:

GO TO SLEEP, COLLEEN, SYLVIA AND VALERIE

CLOSE YOUR EYES, ALDER, ELLEN AND DEAN

SWEET DREAMS, CHICKIE, REVERE AND LITTLE LEAH

CATCH THE TRAIN TO DREAMLAND, JESSE AND IRIS

AND DON'T FORGET TO BRING ALONG

OUR BRAND-NEW LITTLE BABY

SWEET LITTLE SISTER DARLING KATRINA

Maurice finishes the lullaby. VALERIE RUDDICK appears.

VALERIE: My name is Valerie. I'm the daughter of Norma and Maurice Ruddick. I'm ten years old and I'm the third-oldest of my eleven brothers and sisters. Dad said he caught Mom's eye as soon as he met her. Mom said she couldn't resist his trim build, pencil-thin mustache, and his pomaded hair parted in the middle! *(She giggles.)* Mom said Dad always courted her like a gentleman. Always showing up at her front door looking like Mr. Hollywood. *(She giggles.)* Some people thought Dad looked like an odd duck. Mom said it was mostly the men, jealous of his good looks and confidence. Mom would say—

NORMA: Call him what you want, but he's a proud man of colour!

VALERIE: Dad always said it was important to keep your head up and work harder than the next person. *(Leaning in and whispering.)* Some women would chase after Daddy. Always flirting and winking at him. Dad would show up to supper dances wearing a green suit jacket and red pants and Mom would have to shoo women off with a stick.

NORMA: Get your hands off him! He hasn't worked hard enough to win me over, let alone the likes of you!

MAURICE: Yup, when Norma and me first married, we cooked a big turkey and wished on the wishbone of the turkey — a lot of people wish on it, eh? I said, Come on, babe, the turkey's dry. Let's wish on the wishbone! ... So, I closed my eyes. Pulled the biggest piece of the bone. *(Beat.)* Norma asked me what I wished for and I said I wished we'd have thirteen kids.

NORMA: You what?! We're near broke, just got married, can hardly afford the roof we're living under and you coulda wished for anything, but you chose a baker's dozen?! I thought you wanted to be a musician?

MAURICE: *(Laughing.)* And I did, I did! *(Beat.)* But, sometimes life happens to you.

WAY WAY DOWN WITH A SHOVEL AND A PICK

I'LL BE DIGGIN' COAL 'TIL I'M OLD AND SICK

AND I TELL MY KIDS "I DO IT ALL FOR YOU"

"IF I DIG COAL, YOU WON'T HAVE TO DIG COAL TOO"

My kids would come home from school, just in time to see me off to work. My daughter Valerie quickly set her bags on the kitchen table and started making me a hot tea and sandwich. She filled my lunch can with a honey-banana sandwich and set my boots by the door.

(Addressing the kids.) Okay, kids! Come and see your daddy off to work now. *(Wrangling his kids.)* Come here you.... Oh, where you going? ...Get over here... Oh, I love you... Oh now Dean, don't you worry... Daddy's gonna be home soon...

As I made my way out the back porch I took one last smell of my wife's fresh bread and potato stew. It was a beautiful Indian summer, as the brown and rusty red leaves danced around Norma's feet. She was taking the laundry down as I reached in to kiss her neck...

Scene Three

NORMA is taking sheets off the clothesline as MAURICE sneaks up behind her to kiss her neck while grabbing her waist.

NORMA: Maurice! *(Slapping his hand away.)* There's no time for that. *(Flirtatiously.)* Besides, it's not Saturday. All right, off to work with ya. Have a good shift. See you tomorrow. *(Over her shoulder, while folding laundry and placing in basket.)* Valerie, fetch me the good thread and put it on the kitchen table, beside the bassinet. Colleen! Get your sisters ready for bathing. Come here, Dean. *(Holding a bed sheet.)* Hold this end for me. Hold it now. Both corners, like I showed ya. *(Looking at Dean intently.)* You look so much like your father. *(Looking out to MAURICE.)* Look at him... Listen...Whistling away like there isn't a care in the world. *(Grabbing the folded end from Dean.)* Thank you, honey. *(To herself.)* I don't know who's crazier; the man who goes into the mine, or the wife who's senseless enough to marry him. *(To MAURICE.)* It takes a crazy person to go down there! *(Grabbing another sheet from the clothesline, while calling out to Iris.)* Iris, don't slouch. Fetch me your Sunday best and I'll mend it in the kitchen. Dress... Oh, heavens! Here, Alder, take this. Dean, show him how. *(Calling out to MAURICE.)* Maurice!? Maurice! I'm wearing Louisa's dress on Saturday night. Saturday night! I need you to pick it up. It's at Louisa's... He can't hear me. Oh, Saturday night! I cannot wait for Saturday night!

She picks up laundry basket and pushes open screen door with her behind. Music is playing on the radio.

Oh, I like this music. Turn that up, Valerie. *(Sits and puts on glasses and begins to sew.)* I never told anyone this; he thinks he caught my eye first. But, I heard your father singing before he had his sights on me. Heard him in church when I was still in high school. He got everybody singing. Even me. I told you about the Liar's Bench, didn't I, Valerie? No?! Well, your father would take me up there and tell me lie after lie. He once told me he was the guitar player for Nat King Cole. *(Scoffing.)* Nat King Cole. *(Sighs.)* Oh, I just love Saturday nights with your father. Putting on "Pretty Peach" Avon and dancing at the Miners' Hall.

She gets up and begins to sing while attempting to dance with VALERIE.

I WATCH HIM EVERY DAY

'TIL HE'S OUT OF SIGHT

I WAVE AS HE GOES WHISTLING DOWN THE STREET

AND THOUGH IT'S SUPERSTITIOUS

I NEVER DO THE DISHES

'TIL THE WHISTLE BLOWS AND THEY ARE SAFELY DOWN BELOW

AND SOMEHOW I KNOW

HE'LL BE AT THE DOOR

SINGING LIKE EVERY NIGHT BEFORE

DON'T ASK ME WHY, DON'T ASK ME HOW

I'LL TELL YOU IT'S JUST SO, SOMEHOW I....

Turn that off, Valerie. *(Radio fades.)* Go upstairs and do your homework.

After VALERIE exits, NORMA picks up KATRINA and cradles her in her arms. She continues to sing.

I KNOW ONE OTHER THING

I KNOW IT WELL INDEED

GOD WOULD NEVER TAKE A MAN WITH TWELVE KIDS TO FEED

Scene Four

Meanwhile, MAURICE is humming while walking to work.

MAURICE: *(Waving to a neighbour.)* Morning, Millie! Say, did Frank fix your car yesterday?

(Waiting for a response. Aside.) Millie doesn't care much for talking. At least to me she don't. Known her for ten years and reckon I still can't get a smile out of her.

(To MILLIE.) Well, tell him if he doesn't get at it by tomorrow, I'm coming over there myself to fix it!

(Waiting for a response. Aside.) ...Oh well. One of these days...

(Continues walking. He spots another neighbour and starts waving.) Hey, William, great game last week, baby! Way to knock it outta the ball park.

(Aside.) He just graduated high school. He'll go straight to the big leagues, I tell ya!

MAURICE enters Cumberland Mine No. 2.

Here we are. The No. 2 mine. Now, the Lamp cabin is where us miners file in at the beginning of our shift. *(A shift.)* One hundred and seventy-four men reported to work that day. *(Beat.)* We change in and out of our work clothes at The Wash House. It's like the locker room before a big game. Yup. We're all a team here. Always chattin'. Talking about our last shift or hunting trip. Chewing the fat, you might say. Doug Jewkes was a young fella with a long face and blackened teeth. He talked a lot. But I didn't mind.

DOUG: Yeah, I don't like the No. 2. She's an old mine. Not sure how many more years she can give us. We're told to set up the timber walls in straight lines 'stead of staggering 'em. It's thinning out the rock. But the company don't care. They just shove us back down there and tell us to dig.

MAURICE: Dougie was always complaining about them stone and lumber pillars — or what us miners call "packs." See, packs are the walls that we build to keep the tunnels sturdy as we dig. Frank Hunter came walking in the Wash House as if on cue.

FRANK: For the love of Christmas, Dougie! You want the boss to hear ya?! What are you going on about?!

DOUG: I'm talking about the No. 2 mine. She's not going to hold much longer. The old-timers know she's not happy no more. I ain't never liked the No. 2 like I did the No. 4 — and it's even worse now.

FRANK: Well, none of us are too proud of the way we've set up them walls, lining them up straight. But, not all "bumps" are bad. Heck. If you're still alive after you hear a "bump" you know it's keeping the coal loose. Ain't that right, boys?...

DOUG: Well, if I had the brains and education like some other fellers, I'd quit!

FRANK: Well, why don't you get some brains and quit already!?

DOUG: Well jeez, Frank...What else am I going to do? What kind of job would I get that's better than this?

MAURICE: I said nothing. I just listened to Frank and Dougie go on. Wore a smile like I always did. Old man Percy Rector was huddled in the corner.

(To PERCY.) Hey Percy, you know the No. 2. What do you think of her?

(Aside.) Old man Percy Rector was admired by most of us miners for his deep knowledge of coal mining. He'd been in the No. 2 mine for twenty years. Percy had this natural talent in which he could pry off a chunk of rock and discover coal with very little effort. He had all the secrets. He knew all the tricks to surviving the mine. He was a burly man who loved his chewing tobacco and spoke very few words.

PERCY: Aye, she's an old mine, boyos. But when the big one comes... She'll take us all when she goes.

MAURICE: And that was that. He just walked out leaving me with Dougie and Frank bickering about Lord knows what.

A whistle blows.

Scene Five

MAURICE: Each day we hop in a trolley. It takes about twenty minutes to make it down to the first level. Hot dog! It's always a treat to find out who I was working with.

(To the men.) Working a double, Dougie?... Frank!... Pep Pepperdine! Hey!... Percy. Currie. Let's see... Myself, that's six... *(Looking around.)* And Garnie... Hurry up, Garnie, you're gonna be late! That's seven. Seven of us would pile into the trolley and make our way down to the centre of the earth. Down the shaft we go!

Trolley begins to descend. There is a long pause.

Twenty minutes.... It wasn't always a pleasant ride. Sometimes it seemed to last a lifetime.

Pulls spoons out from his lunch box.

So, I liked to pass the time with a song or two. You know, start the day right...

Begins to sing:

NUMBER TWO MINE IN CUMBERLAND COUNTY

FILLING OUR HANDS WITH THE GOOD LORD'S BOUNTY

TURNING COLD DARK NIGHT INTO SUNNY GOLD

WORKING ON OUR KNEES 'TIL WE'RE SICK AND OLD

SWINGING A PICK WHERE THE SUN DON'T SHINE

WHERE'S THAT, BOYS?...

CUMBERLAND COUNTY DOWN NUMBER TWO MINE

OLD NUMBER TWO IN SPRINGHILL TOWN

TOUGHEST MINERS FOR MILES AROUND

YOU'LL SCRATCH AT THE COAL 'TIL YOUR FINGERS BLEED

WITH A HOUSE FULL OF HUNGRY BELLIES TO FEED

POCKETS ARE EMPTY, WHAT DO YOU DO?

TELL 'EM, BOYS

GO TO SPRINGHILL TOWN DOWN OLD NUMBER TWO

It's blacker than coal down here. So black, they named a colour after it. Coal-black. There's youngsters down here who dream of bigger things. Better things. The older boys — the men who've been working down here for most of their lives know they're in it for the long haul. Me?... I'll probably die down here working in the mine...

Sings.

NUMBER TWO MINE IS ONE GLORIOUS HOLE

AND CUMBERLAND RAILROAD OWNS MY SOUL

THEY'RE THE ONES GURANTEE MY PAY

I PLAN TO GET RICH ELEVEN DOLLARS A DAY

ELEVEN DOLLARS A DAY, TELL ME, WHERE DO I SIGN?

SHOW 'EM, FELLAS

CUMBERLAND COUNTY DOWN NUMBER TWO MINE

MAURICE exits the trolley.

It was pretty clear what our job was. Clear as mud. Dig the tunnel for coal. We shovel sixteen tons for eleven dollars a day. Once you run out of battery power you won't be able to see your own fingers in front of your face. Back in the day, before I was a miner — even before my father worked in the mines — they had mules dragging out the coal and the miners used kerosene lamps. Lamps with flames. That's crazy! You see, mining is one of the most dangerous jobs. The toxic gases alone down here is enough to blow you to Kalamazoo.

Begins to work and sing:

OH, THE LIAR'S BENCH IS A MIGHTY FINE BENCH

AND A LOVELY PLACE TO SIT

SOME GO THERE TO SPIN A YARN

OTHERS LIKE TO SHOVEL THE...

COAL, BOYS! COAL, BOYS!

IT'S HOW WE EARN OUR NICKEL...

It's hotter than July down here. The air fans don't help much neither. Yeah, nothing stays cool down here for long. You can't escape the sound of the coal rakes. The picking of rock slaps at ya. The sweat. The dampness. Yup. This is what it's like working in the mines. You're in the middle of it. Constant. Eight hours a day. *(Imitating NORMA.)* "You have to be crazy to work in the mines." *(Smiling. Aside.)* That's what our wives always tell us. "It takes a crazy person to go down there." Well, you can see the row of us. All half-witted and fearless. Working just as hard as me. Faces blackened by the dirt and coal. Faces, just as black as me.

A whistle blows.

On my breaks, I usually ate alone. But on this day I thought I'd join Percy for lunch.

(To PERCY.) Mind if I join you, Percy?

(Aside.) There weren't many workers beneath Springhill I considered calling good friends. Oh, they were decent enough. But old man Percy... There was very little sunlight in his world. It seemed as the years dragged by, so did his thirst for living.

PERCY: *(To MAURICE.)* You're the same as us, I figure. Never knew a nig... a coloured boy before I met you. You're a good worker, boyo. And don't mind us callin' you "boyo" or nothin' neither. Don't no one mean nuthin' by it. Just up there, we don't eat together. Mind you, I ain't never been invited to the bosses' table neither. Yup. Down here. We're all miners. We're all the same colour. If you really thought about it. A miner's job ain't fit for no man.

Pause.

MAURICE: *(To PERCY.)* Look at this. My ten-year-old daughter — Valerie — made me a gourmet sandwich. Now, look at the honey oozing outta the corners. Doesn't that look scrumptious? Would you like a piece, Percy?...

MAURICE hands him a sandwich. PERCY offers MAURICE a stick of gum.

A stick of gum. Don't mind if I do. I haven't had gum since—

MAURICE is startled by a small bump.

Holy Jinglan. Was that what I thought it was? ....

PERCY: That *(Spits in a can.)* was a bump.

MAURICE: Oh. Cripes.

(Aside.) A miner knows a bump. Some would call it a tremor — a quake. To us miners it felt more like a bump. A flutter in the walls. A small bump happened once or twice a month. This was usually a good thing. Meant it loosened up the coal, made it easier for digging.

(To PERCY.) But, if you hear a bump, you're okay? Is that right, Percy?...

PERCY stands.

It's the bump ya don't hear that kills ya.

He exits.

MAURICE: And that was that. Percy just walked out. Into the darkness.

(To PERCY.) Thanks for the gum, Percy. You take it easy. See you at the top.

A whistle blows.

MAURICE starts to work.

Oh well, you gotta make a living. After all, that's what makes the bacon sizzle. *(Chuckles to himself.)* Bacon sizzle...

Continues to work as he sings:

KEEP ON DIGGING

KEEP ON DIGGING

IT'S WHAT US MINERS CALL A LIVING

PICK AND SHOVEL, HAMMER AND CHISEL

THAT'S WHAT MAKES THE BACON SIZZLE

MACHINES KEEP DRUMMING

THE WORLD KEEPS HUMMING

SURE AS THEM BABIES KEEP A-COMING

SURE AS SUNSHINE, SNOW AND DRIZZLE

YOU KNOW WHAT MAKES THE BACON SIZZLE

AND SHOULD I EVER BREAK FREE OF THESE EARTHLY BANDS

I'LL SHOW ST. PETER MY CALLOUSED HANDS

THE BACK I BROKE AND THE MUSCLES I BURNED

A RECORD OF EVERY LAST CENT I EARNED

SO QUIT YOUR BAWLING

KEEP ON CRAWLING

LISTEN TO THE SOUND OF YOUR BABIES CALLING

YOU CAN'T QUIT 'TIL YOU'RE OLD AND GRIZZLED

'CAUSE THAT'S WHAT MAKES THE BACON SIZZLE

YES, THAT'S WHAT MAKES—

Singing is interrupted by a louder bump. THE bump. Lights out.

Scene Six

Cumberland mine. JACK ad libs: "How's my tie?...Mic check...Are you getting a signal?...Three... two..." CBC theme music is heard.

JACK: This is Jack McNeil reporting for CBC. At 8:06 pm the town of Springhill experienced a "bump." The "bump" was not a cave-in. The rock floor rose upwards. The pressure of the stone two miles beneath the surface caused the coal to erupt. The stone floor at 13,400 feet shot straight up and crashed into the 13,000-foot level. The stone and lumber pillars were shattered in an instant. The ground shook over a fifteen-mile radius. Only moments before, families were nestled in front of TV sets watching the *Howdy Doody Show* and *Don Messer's Jubilee...*

Don Messer's song "Further Along" is heard on a distant television.

VALERIE appears.

VALERIE: We were sitting in front of the TV, singing along in our pajamas *(Stops singing and looks down.)* I felt the floor being lifted. It was like a bang. A huge thump! *(Beat.)* The "bump." The TV went all white. Shwoosh! We all heard stories. But, Mom knew exactly what a bump was. She was so tired from having our little sister Katrina, only two weeks old. But, she jumped up like the house was on fire. She ran through the kitchen to check on my brothers upstairs. Next thing she knows she's juggling coffee mugs as they bounced off the cabinets. When she came back into the living room, me and my two older sisters — Colleen and Sylvia — were sitting on the edge of the couch, wearing coats and scarves. We were just looking at her. Waiting for her to tell us what to do. *(Beat.)* She had this look in her eyes I ain't never seen before. It got me all scared. Worried. All she could say was...

NORMA: Go to the mine. Go get your father.

Minutes after the bump. Two miles below the surface, in the Cumberland No. 2 mine. MAURICE is crawling on his hands and knees in the dark. There is screaming in the distance.

MAURICE: Who's screaming?!.... Where's my hat?.... Oh, dear Lord. Can't lose the light. *(Hears another voice.)* Hello? *(Beat.)* Who's there? *(Beat.)* Garnie? *(Beat.)* Garnet Clarke?! Is that you?! *(Beat.)* Yeah, it's me. Maurice. Follow my voice. Garnie, over here. Garnie *(Beat.)* When was Currie Smith last with ya, Garnie?...Okay. Who else is with ya?... Frank?! *(Noticing Frank's head wound.)* Cripes. *(Doug Jewkes appears.)* Atta boy, Dougie...*(Losing breath.)* I heard that "bump," then everything went spinning. One minute I'm hanging the timber boards, next thing I know I'm buried in coal and rock. I was buried to my waist, boys! Twisted like a bent screw. *(Hears shuffling and moaning. Calls out.)* Pep?... That you?..*(PEP appears.)* Oh, cripes. Who's screaming? Who's here? Let's do a count? I have some chalk. Let's mark our names down on this timber — don't talk like that, boys. We're going to make it out alive. The gas isn't gonna kill us. Just as long as we keep our heads down and our wits about us. Never mind that! We're going to get out of here alive! *(Finds a piece of chalk.)* Let's see, there's me. Pep. Currie. Garnet and Frank and Doug. *(Looking around.)* Okay, well—who's screaming? Don't try to stand up, Frank. Looks like we only got three feet above us. *(Beat.)* I'm going to crawl around and follow that scream.

(Begins to crawl.) I'm heading towards you. I hear you. I'm crawling towards you! Calm down. Let me get the light on you. Oh. No. Percy?... That you, Percy? *(Beat.)* Oh, cripes...

(Aside.) Percy was hanging from the wall. His arm wedged between a timber pack and a wall of coal. His feet were dangling above the ground. I tried to pry his arm out from behind the rock. Percy screamed in agony. To hear this deep scream from a grown man was gut wrenching. Not a good sound. I tried to tuck myself under Percy's body, tried to keep him raised. Relieve some of his body weight. Everyone started talking at once.

(To the men.) Calm down! Quiet, boys. We're louder than Percy. Garnie, go look for a jack out on the timber road. Pep and Frank, go look for another exit. No, Frank. We're not going to cut his arm off. Watch your battery pack, Dougie... Make sure you have enough light, Currie. *(To PERCY.)* Now Percy, I need you to breathe, babe. We're going to get you out of this. Take some water. Slowly now. There now.

(Aside.) Percy kept on squirming. Like one of them possums trapped in a cage. Frantic and all.

MAURICE begins to sing:

KEEP ON DIGGING

KEEP ON DIGGING

IT'S WHAT US MINERS CALL A LIVING...

Lights fade on MAURICE.

Scene Seven

Cumberland mine. CBC theme music is heard. Lights up on JACK MCNEIL.

JACK: This is Jack McNeil reporting to you live in Springhill at the Cumberland Mine No. 2 mine at the pithead. I have just received word that rescue workers have been called in from Halifax, New Glasgow and the surrounding area to dig for any survivors. In the first day since the "bump," small groups of miners have surfaced. Many have been sent directly to the hospital. It is a gruesome display of wounded men. The rescue workers believe it will take more than a week to find any bodies amongst the two miles of rock. With that distance, there is no doubt we will be here for quite a while. Until then, stay tuned. Jack McNeil reporting for CBC, Springhill.

Lights fade on JACK.

MAURICE appears out of the darkness.

MAURICE: We think it was a Saturday. Day three since the "bump." There were seven of us. Stuck in a dark hole. Two miles beneath Springhill. Even though we had very little food and water, everyone seemed all right. *(Beat.)* Everyone but Percy. We were trying to make sense of it all. Frank was having a hard time hearing. And it was getting worse by the minute.

FRANK: When the bump took me down I noticed me hat was gone, and that's when I felt something wet around my ears. Dougie, can you see anything around my ears?...Why are you moving your mouth and not saying anything, Dougie?...Do you boys hear that ringing?

MAURICE: Frank had gone deaf.

(To FRANK.) Just sit quietly, Frank.

(Aside.) I noticed Pep in the corner. His arms wrapped around his legs, rocking back and forth.

(To PEP.) Pep, are you okay?

(Aside.) He just gestured towards his feet as I crawled closer. It was his brother-in-law. *(Beat.)* Laying there. Silent and still.

PEP: We used to go fishing together, Maurice. Ain't no way I know how to tell my sister her husband is gone. That's my brother-in-law there, Maurice. That's my brother-in-law...

MAURICE: Pep just kept on repeating those words as he rocked back and forth. It was no use trying to calm him down. And it was no use trying to sleep. My headlamp was fading and my battery pack was running low. I didn't want to be trapped down here in the dark. It's one thing being trapped. But, in the dark...This type of darkness can drive a man insane. You don't know true darkness until you been down in the mine. You can shut yourself in a closet and put a towel in the cracks of light, but that ain't nothing compared to this darkness. So, Currie, Garnie, Frank and me started crawling for tools and battery packs. *(Beat.)* All the bodies...Lifeless. Cold bodies. Everywhere. We started patting them down. Searching for food, water canteens. Anything. Currie Smith came out of nowhere, excited as all get-up.

CURRIE: Look it here, boys! I struck gold! I tell ya, I struck it rich. A chocolate bar! I found a Milky Way — not even opened — found it in that there fella's front pocket. It's all right, isn't it, boys? Now, let's see here. We'll split it six ways.

MAURICE: Split it seven ways, Currie. I know he's not all there, but Percy deserves a piece just like the rest of us. I'll take two. One for him and one for me.

He crawls to PERCY and hands him a piece of chocolate.

Here you go, Percy. Try this tasty bit. Now, take it easy. Chew it slowly. In fact, don't even chew it at all. Let it just sit there and melt on your tongue. There now. You'll be okay. *(Distant sound of chipping.)* Wait...What's that sound?... Quiet, boys! I need to hear. No, Frank. There ain't no beer! Listen. *(Hears distant chipping.)* Yeah. Those are rescue workers. They're coming down to rescue us! It won't be long now, boys. Sit tight. Save your energy.

Pause.

Minutes...hours dragged by. We started using Percy's screams of agony as the second hand on a clock.

(To PERCY.) Ssh. There now, Percy. It'll be all right. I need you to keep quiet. *(To the men.)* Get your ears against those planks, boys. Keep listening for the sound to come closer. Those are trolleys and drills. Let's hope they know we're here. *(Calling out.)* Hey there! There's seven of us. Help!...

Lights fade.

Scene Eight

CBC theme music is heard. Lights up on JACK.

JACK: It's day four since the largest accident in North America's deepest coal mine. Only small groups of men have been brought to the surface, bringing the count of survivors rescued to ninety-three. However, eighty-one men are still unaccounted for. And, they have no definite indication that any of the remaining eighty-one are still alive. Mine rescue crews are being hampered in their rescue efforts by huge pockets of methane gas, fallen timber and coal. Many wives spend all night at the pithead. They wait, huddled against buildings, hoping against hope that there are any survivors. Personally, I don't see any of those Nova Scotia miners coming out alive. This is Jack McNeil of the CBC, reporting for Springhill.

Lights fade on JACK.

Cumberland Mine No. 2. Two miles below. Day four. Darkness.

GARNET: Boys?... Boys, it's Garnie. Wake up. It's Monday. Today's my birthday. Maurice, wake up. It's my birthday. I'm twenty-nine. Would you believe it? It's my birthday, and I'm spending it with the likes of you fellas! The way it looks...I won't get any older.

MAURICE: Most of us had surrendered to our fate. But, the thought of celebrating a birthday made me wanna sit up and look forward to something, anything but death.

(To the men.) Let's have a birthday party for Garnie! Come on boys, wake up.

(Aside.) Currie sat up and chuckled...

CURRIE: Ain't no way I'm going to a party without a shave. Better get ready for a party. Anyone got a razor? Better get ready for a party now. Ain't no way I'm going to a party looking like this.

MAURICE: *(To the men.)* There's half a sandwich left. My daughter Valerie made it for me. I think it's time we split it up. Now, let's have some water. A cap of water... *(Smiling.)* A glass of champagne. Let's have a glass of champagne. *(Whispering.)* I know it's not champagne, boys, but let's just pretend. Well, here's to Garnie, boys. Come on, let's hear ya!

Prompting them to sing:

...HAPPY BIRTHDAY, DEAR GARNIE

HAPPY BIRTHDAY TO YOU...

Continues to sing.

OHHHH....

THE LIAR'S BENCH IS A MIGHTY FINE BENCH

AND A LOVELY PLACE TO SIT

SOME GO THERE TO SPIN A YARN

OTHERS LIKE TO SHOVEL THE

COAL, BOYS; COAL, BOYS

IT'S HOW WE EARN OUR NICKEL

YOUNG GARNIE'S GOT THE BIGGEST PICK

BUT HE DON'T GOT THE BIGGEST...

PAY CHEQUE, PAY CHEQUE

THREE CHEERS FOR THE WORKING CLASS

CUMBERLAND COUNTY DON'T OWN MY SOUL

AND IT CAN KISS MY...

ASK THE BOSS FOR A NICE BIG RAISE

HE'S CERTAINLY GOT THE RICHES

AS IF THEY'D PART WITH ONE THIN DIME

THE DIRTY SONS OF BRITCHES

...AND RUBBER BOOTS...

MAURICE hears PERCY moaning.

Our party was interrupted by Percy. He could barely breathe as he pulled me in closer with his only good arm. With all the energy he had left, he leaned into me, saying—

PERCY: Cut it off, Maurice. Cut off my arm. Leave it buried in the No. 2. I won't blame you. I want to go home. I want to go to my swimming hole with Lucy. It's so cold. The water is so cold.... Lucy?...

MAURICE: I leaned towards him and held his hand as he choked his last breath. *(Pause.)* And... That was that. *(Pause.)* Poor Percy. There was just six of us. Darkness fell on us like no other darkness had fallen on us before. It was as if Percy took a piece of each one of us when he left us. If we were to ever make it out alive, this would be a birthday party Garnie would never forget. *(Pause.)* Time passed. We lost track of how many days. *(Beat.)* My mind started to rot. *(Beat.)* I couldn't remember...

(Shaking.) I have ten kids. No. No! Twelve....

While struggling to remember, he sings:

GO TO SLEEP, COLLEEN, SYLVIA AND VALERIE

CLOSE YOUR EYES, ALDER, ELLEN AND DEAN

SWEET DREAMS, CHICKIE, REVERE AND LITTLE LEAH

CATCH THE TRAIN TO DREAMLAND, JESSE AND IRIS

AND DON'T FORGET TO BRING ALONG OUR BRAND-NEW LITTLE BABY

SWEET LITTLE SISTER DARLING KATRINA

HERE'S A KISS GOODNIGHT FROM YOUR DARLING PAPA

HE'S GOING OFF TO WORK DOWN IN THE MINE

PROMISE YOU'LL BE GOOD FOR YOUR DARLING MAMA

SWEET DARLING NORMA

TELL HER PAPA'S FINE

Lights fade on MAURICE.

Lights up on VALERIE. She is either pretending, or she is strumming a real guitar while singing. The author leaves it up to the actor/ director to decide.

VALERIE: REACH RIGHT UP AND KISS THE MOON

VALERIE'S LOVING DADDY'S GONNA BE HOME SOON

SHE'S SWEET AS HONEY IN A SPOON

AND VALERIE'S LOVING DADDY'S GONNA BE HOME SOON

(Looking up.) I hope you can hear me. It's been seven days since you've been gone. We missed you at church on Sunday. We sang your favourite hymn.

She begins to sing "Rock of Ages."

ROCK OF AGES/ CLEFT WITH THEE...

She tries to remember the lyrics. Frustrated, she stops singing and clasps her hands together as if she is wanting to pray.

Don't you dare leave me alone with my little brothers and sisters. You've got a two-week-old and Ellen just got over the measles. Mom and you worked too hard to make this family what it is now. Don't leave it all behind without at least saying goodbye. You still have so much more to show us in this world. Leah's been practising that song you taught her. Alder still wants to go fishing and Revere wants to ride trains and sing songs. The TV and newspapers keep telling Mom to get ready for the worst. But, she ain't ready. So, Daddy...come home soon.

Lights fade on VALERIE.

Scene Nine

Cumberland Mine No. 2. Two miles underground. Eight days since the bump. Darkness.

MAURICE is exhausted and fading. A choir is heard in the distance, singing "Rock of Ages." MAURICE sits up and tries to sing along.

MAURICE: ROCK OF AGES, CLEFT FOR ME

LET ME HIDE MYSELF IN THEE

WHILE I DRAW THIS FLEETING BREATH

WHEN MINE EYES SHALL CLOSE IN DEATH...

MAURICE struggles for air. There is silence as he sits in darkness. He collects himself and begins to sing with more energy.

A MINER'S MADE OF SWEAT

A LUNG FULL OF AIR AND A BEATING HEART

ONE OTHER THING TO THANK GOD FOR

AND IT'S THE MOST IMPORTANT PART

THE THING THAT'S FORGED INSIDE OF US

AND IN THE SOULS OF THE FOLK ABOVE

IT'S A JEWEL THAT WILL LAST THE AGES

THIS JEWEL CALLED LOVE

THOUGH OUR HEADLAMPS NOW ARE FADING

AND SOON IT WILL BE OUR BREATH

THERE'S A DIAMOND SPARKLING IN EACH OF US

THAT WILL OUTLAST DEATH

ALTHOUGH OUR TIES TO ONE ANOTHER

TIME WILL SURELY SEVER

OUR LOVE IS A LIGHT THAT WON'T GO OUT

IT WILL SHINE FOREVER

IT WILL SHINE FOREVER

IT WILL SHINE FOREVER

OUR LOVE IS A LIGHT THAT WON'T GO OUT

IT WILL SHINE FOREVER

MAURICE stops singing. He hears PEP in the corner.

Pep?... Pep, what's wrong?

PEP: I have three kids and a loving wife back at home, Maurice. Would you believe it? This is the farthest I've been from home and I'm two miles beneath Springhill.

MAURICE: *(Aside.)* We had no food left. No water. We chewed on coal and ate timber bark. So, I suggested to the boys we drink our own...

MAURICE turns his back to the audience, unzips his fly and urinates in a cup. He then faces the audience, smells the cup full of urine and reluctantly drinks it.

It was disgusting — we had no choice. We were dying of starvation. Dougie sat up against the stone wall, shaking from weakness.

DOUG: You know what I keep thinking about? I can't help it, but do you know what I keep thinking about?... 7 Up floats. Back at home, June and me — we used to make 7 Up floats. If I ever get out of here... I'm going to make me a giant glass of 7 Up float.

MAURICE: That's a good idea, Doug. Think of this as a tall cold glass of 7 Up float. Cheers, boys. Here's to a tall cold glass of... 7 Up float.

He passes the cup to FRANK.

FRANK: You still got high hopes of getting out of here, Maurice? You still think we're going to make it out alive, eh? Well, you're dumber than you look. What gives, anyway? You're just some odd-looking coloured that chirps like a twisted canary. And now, you got us eating coal and drinking our own piss... Well, I'm sick of it. I'm sick of the lotta yas! And, Dougie — if you don't move your feet, I'm going to knock what teeth you have left in that fat mouth of yours. Try sipping on 7 Up after that!

MAURICE: We were near the end, and we all knew it. It was just a matter of time when we would be joining Percy. I found I was digging deep into myself to find that faith that was once strong when I was a child, but now seemed so distant.

He sings:

OUR LOVE IS A LIGHT THAT WON'T GO OUT

IT WILL SHINE FOREVER...

Lights fade on MAURICE.

Scene Ten

CBC theme music. Lights up on JACK MCNEIL.

JACK: This is Jack McNeil for CBC News. The day is Friday, October 31st. Halloween night. Eight days since the mining disaster. The last word — the last informed word — is that it would take a miracle for any human being to survive the depths of the mine. The lack of oxygen, food, and water would make it improbable after eight days. As I report this to you, there are volunteers bringing out what is expected to be the last survivors of the mine. We can only assume the gratitude and relief the miners' wives and kids must feel. Sadly, we send our condolences out to the families of those who are presumed dead. This is Jack McNeil...

He notices NORMA. She is with VALERIE, REVERE, LEAH and JESSE.

Excuse me. Excuse me, Miss... What's your name?

NORMA: Norma. And, it's Missus. Mrs. Maurice Ruddick.

JACK: Those are very sweet children. Our condolences to you.

NORMA: What? There's no need for sympathy. My Maurice is coming home soon.

JACK: So, you hold on to faith that your husband is still alive?

NORMA: Of course I do.

JACK: Have you lost other family members in the mine?

NORMA: Well, no. But our neighbour's husband didn't make it out alive, and Valerie's friend at school lost her grandfather in the mine. The minister came to their house yesterday.

JACK: Mrs. Ruddick... Have you thought about what you might do if your husband doesn't come home?

NORMA: Valerie... Go take the kids and get Mama some coffee. Go on.

She watches VALERIE and the kids exit. Then turns to JACK.

Now you listen to me, Mister McNeil. Maurice is gonna be home soon. I just know it. Sure, it scares me to think about it. But, I know I have to be strong for my kids. I have to be strong for my Maurice. I know he'd want it this way. So, if you'll excuse me, sir...We have some praying to do.

Norma exits as Jack is left speechless. He looks back to the camera.

JACK: ...This is Jack McNeil... CBC News... Springhill, Nova Scotia.

Scene Eleven

Cumberland No. 2 Mine. Eight days since the bump.

MAURICE crawls out of the darkness. He is exhausted and barely conscious. He sings:

MAURICE: TRAPPED IN THE DARK TWO MILES UNDERGROUND

TWO MILES BENEATH THE GOOD FOLK OF SPRINGHILL TOWN

OH DEAR LORD, HEAR THIS MINER'S PRAYER

IS THERE ANYONE THERE?

IS THERE ANYONE THERE?

(Looking up.) Please, God. Help me. Say this isn't the end. Tell me I'll see my wife. My kids. My freedom.

(Aside.) I could hear the heartbeats of the men beside me fade. Weaker and softer they became. I could hear whimpering in the corner. Muttering.

(Calls out.) Who's there?... It was Pep. He was repeating two words over and over again: "Why me? Why me?..." Then it hit me. Why not me? Why not us? We don't get to choose our own way of dying. Maybe God figured since we were born down in this mine we'd die down here. We ain't done anything else but build these dark alleys.

EVERY NIGHT ON MY KNEES I THANK YOU FOR MY NORMA

AND THE BABIES THAT YOU BLESSED

AS I TAKE MY REST I THANK YOU FOR ANOTHER GLORIOUS DAY

BUT NOW I'M DONE, I'VE REACHED THE END

AND DAYS THAT SEEMED TO CARRY ON

FROM DAWN UNTIL THE DUSK OF TIME

ARE RUNNING DOWN AS SURELY THEY WERE MEANT TO DO

SO HERE I AM

I'M ON MY KNEES

I THANKED YOU FOR ALL THOSE DAYS

I GUESS THERE'S NOTHING LEFT TO DO

BUT THANK YOU, GOD, FOR THIS DAY TOO

He collapses from exhaustion. Knowing he may never wake up, he begins to fall asleep.

GO TO SLEEP, COLLEEN, SYLVIA AND VALERIE...

Fading.

REACH RIGHT UP AND KISS THE MOON

VALERIE'S LOVING DADDY'S GONNA BE HOME SOON...

Fading.

WAY WAY DOWN DIGGIN' DOWN IN THE DEEP

I'M A COAL-DIGGIN' DADDY DIGGIN' COAL FOR MY KEEP...

Diggin' ... Diggin'...

Silence.

MAURICE suddenly wakes up.

No. No. No. This isn't right. I ain't givin' up. *(Beat.)* Come on, boys! Wake up! Listen to me. We ain't going down like this.

Gathering his strength while singing:

IF I COULD SING JUST ONE MORE SONG

GIVE ME THE STRENGTH TO CARRY ON

IF YOU GIVE ME JUST A FEW MORE DAYS

THEN I'LL FILL THE WORLD WITH THANKS AND PRAISE

IF I COULD SING JUST ONE MORE SONG

GET MY FRIENDS TO SING ALONG

IF YOU LET US MAKE A GLORIOUS SOUND

MAYBE WE WILL WALK AGAIN ON SOLID GROUND

(Rising to his feet.) I ain't giving up until we all get out of here alive. And then, I ain't never coming back.

He crawls for his lunch box and pulls out his spoons.

I started banging on an air pipe. Three short, three long, three short. I stayed at it for a half hour, then an hour. Then when I was too tired to go on, Garnie would take over. Then Currie. Then Pep. *(Continues tapping.)* Just then... A faint distant sound.

Quiet, boys!

Beat.

There was a tap. It was distant, but I know it was there.

(To the boys.) Did you hear that? Was it just me or did you hear that?!

There was silence. *(Pause.)* But, only for a moment.

Chipping and tapping is heard in the distance.

It began again. Tap tap tap. I responded. Three short, three long, three short.

A crack of light pierces through the darkness.

And... That was that. Beams of light flashed through the air pipe as we heard voices from above. It was the rescue workers! They were like angels. Next thing I remember, Garnie gave me a kiss on the forehead.

(Laughing with GARNIE.) Okay, get off me! *(Calling out.)* Hey there, it's Maurice! Maurice Ruddick! There's seven...There's six of us. *(Beat.)* Oh boy, get us some water and I'll sing you a song! Yippee!

Beat.

It took two hours for them to dig through the tunnel. They poured hot coffee and soup down the air pipe and kept us hydrated 'til they could reach us. When the rescue workers opened up the tunnel, we saw Percy's body in the light. *(Beat.)* We weren't fully able to see him until then. *(Beat.)* It was horrible. *(Pause.)* I said my last good-bye to Percy as we crawled our way to the top. It was Saturday. November 1st. 8:45 pm. There was over one hundred news journalists with flashbulbs popping. *(Beat.)* The light! *(Beat.)* We were finally free. *(Beat.)* As I stumbled out of the pithead, I took in the free air. Nine days without proper food and water. We were dying of thirst. Starving. They quickly rushed us to the hospital. Norma and the kids met me there, dressed in their Sunday best. And Valerie...Valerie had a tasty honey-banana sandwich waiting for me. Just the way I liked it. And believe me, it never tasted so good. *(Beat.)* People came to visit us from all over. The Prince of England even flew down! In the United States, the governor of Georgia invited everyone to an all-expense-paid trip to a private resort called Jekyll Island. Everyone was invited. Everyone! Everyone... But me? *(Pause.)* The governor wouldn't allow me to come on account of me being coloured. *(Beat.)* When some of the boys found out about this, they raised a stink and told the press...

FRANK: We ain't going if Maurice ain't going.

MAURICE: So, would you believe it? I ended up coming with my family. They had us stay a mile down the road in a trailer park not far from the resort. The coloured people of the community came out of their homes and greeted us. They ain't never seen an African-Canadian before — a mulatto like me. They treated me like royalty. *(Beat.)* The governor must have changed his mind. He invited all the miners and wives to the big house for a huge feast — including me! As I sat there with the bosses...the governor...Garnie. Frank. Currie. Dougie. Pep!... I couldn't help but think of Percy.

Beat.

(To the boys.) If only Percy could see us now. Eating together at the bosses' table.

(Aside.) It occurred to me just then that we were all equal.

Beat.

The governor asked me when I was going back into the mine. I just looked at him and said; "Well, sir...never." The boys looked at me as if

I had broken the miners' code. But then they too realized they could break it as well. *(Beat.)* And they did. *(Beat.)* Back in Springhill, I heard Garnie took up odd jobs and celebrated his birthdays with his mom at home. Pep learned to play accordion. Frank never was the same after he came out of the mine. He never did get his hearing back, and it turns out he lost his brother down there and he didn't even know it. And, I think he regretted not cutting off Percy's arm. I see Currie once in a while on Main Street. We exchange pleasant glances. Oh, and Dougie?... Hot dog! Dougie got a job in Toronto, working for 7 Up!

Beat.

Me, I started a band with my older girls. We were called "The Singing Minerettes." We'd hop in the car on weekends and sing at supper dances and pool halls. We were quite the hit! *(Beat.)* The premier of Ontario awarded me Citizen of the Year! I was the first African-Canadian ever given the award. Well, if it was up to me, I think all of us miners should have been given an award. We were all heroes down there.

Beat.

I'm still the same man I was before the bump. You ask me what kept me alive down there? Well, it was my faith. My music. My family. *(Beat.)* The thought of feeling the cool summer breeze and sunlight hitting my forehead. Enjoying a tasty honey-banana sandwich. Or singing and writing songs on my guitar.

Beat.

Now, that's living. And, I wouldn't change it for nothing.

He sings:

NUMBER TWO MINE, YOU CAN'T HAVE MY SOUL

'CAUSE I'VE HAD MY FILL OF SCRATCHING AT THE COAL

GONNA SIT ON MY PORCH AND BREATHE THE SWEET NIGHT AIR

AND PRAY FOR THE MEN THAT WE LOST DOWN THERE

LOST DOWN THERE 'TIL THE END OF TIME

CUMBERLAND COUNTY DOWN NUMBER TWO MINE

The End

Beneath Springhill: This Miner's Prayer

Down in the Deep

Somehow I Know

rit.
23 C
D
'Til the whis - tle blows and they are safe - ly down be - low; And
a tempo
25 G
Em7
C
D
some - how I know he'll be at the door,
27 C
Am7
D
Sing - ing like ev - ery night be - fo - re.
29 C
G
Don't ask me why, don't ask me how, I'll tell you it's just
31 D
G
so Some - how I know I
Even eighths
33 C
G
Em7
know one oth - er thing I know it well in - deed
rit.
35 Am7
C D G
God would ne - ver take a man with twelve kids to feed.

Number Two Mine

Rob Fortin

Susan Newman

Number Two Mine

Lyrics by Rob Fortin

(possible call & response)

Number Two Mine in Cumberland County *(repeat)*
Filling our hands with the good Lord's bounty *(repeat)*
(All) Turning coal dark as night into sunny gold
Working on our knees 'till we're sick and old
Swinging a pick where the sun don't shine *(hold note)*
"Where's that, boys?"
Cumberland County down Number Two Mine

Old Number Two under Springhill town
Has got the toughest miners for miles around
You'll scratch at the coal 'til your fingers bleed
With a house full of hungry bellies to feed.
So, if your pockets are empty what do you do
"Tell 'em, boys"
Go to Springhill town down old Number Two

Number Two Mine is one glorious hole
And Cumberland Railroad owns my soul
They're the ones guarantee my pay
And I plan to get rich on eleven dollars a day
Eleven dollars a day! Tell me where do I sign?
"Show 'em, fellas"
Cumberland County down Number Two Mine

Number Two Mine is where I'll live and die
And never see the sun shining in the sky
I'm in the devil's debris from noon to night
But coal provides the people with heat and light
Hop in the trolley, it'll be just fine
"Where's that?"
Cumberland County down Number Two Mine

(2)

Cumberland County down Number Two Mine
Your battery dies you'll be working blind
Blind as a bat in this stinking cave
That any second might be your eternal grave
My time is up, Lord, give me a sign
"You know where to find me."
Cumberland County down Number Two Mine.

Down Number Two with a pick and a drill
Fire don't get you then the gasses will
Most dangerous place on God's green earth
Shovelful of coal's what your life is worth

Maurice's Lullaby

The Liar's Bench

What Makes the Bacon Sizzle

Rob Fortin

Susan Newman

Daddy's Gonna Be Home Soon

Coal Miner's Hymn

For This Day, Too

Maurice Ruddick leaning against his car. From the Ruddick family collection.

Maurice and Norma with their kids in the kitchen of the house on Herrett Road in Springhill,Nova Scotia in 1959. Photo by Bob Brooks.

Norma and kids watching TV. Photo by Bob Brooks.

Maurice receives a plaque from the Canadian government honouring his contributions to his community. From the Ruddick family collection.

Maurice stands at the pit head of the No. 2 Cumberland Mine. From the Ruddick family collection.

Norma and Maurice holding a demo reel of Maurice's original compositions. From the Ruddick family collection.

Beau Dixon with members of the Ruddick family, February 2013. Photo by Doug Wyse.

Beneath Springhill: The Maurice Ruddick Story

Study Guide

compiled by Firebrand Theatre
www.firebrandtheatre.com

Introduction

Maurice Ruddick (1912–1988) was one of seven miners trapped underground in the 1958 Springhill mining disaster in Nova Scotia. He was the only African-Canadian of the seven men who were buried alive that day. Once miraculously freed from the mine, he was celebrated as "the reason the men survived." Ruddick tried to keep spirits high by getting the men to sing hymns, though most were badly injured and almost dead from dehydration, having been stuck underground for almost nine days. In order to survive they had to scavenge for food and water in the clothes of other dead miners. Eventually, they had to resort to drinking their own urine.

Ruddick was a father of twelve at the time of the disaster; his youngest was just two weeks old when the "bump" happened. He would later have one more child, making a total of thirteen. At the time of the disaster, he was 46 years old. He had worked in the Cumberland No. 2 mine for about 20 years, but his last day of work as a miner was the day of the explosion.

After the men were rescued, the six surviving men were sent to stay at the hospital for a few days. Prince Philip flew from England to meet them all. A few of the men were asked to be on the *Ed Sullivan Show* in New York. When Maurice was released from hospital, he was awarded Citizen of the Year by the Ontario premier Leslie Frost. The governor of Georgia heard about their plight and invited the men to an all-expenses-paid trip at a private resort called Jekyll Island. When the governor found out there was a "coloured man" among them, he told the men that Maurice would not be allowed to join them. When the miners heard this they told the press: "We ain't going if Maurice ain't going." Maurice was allowed to stay with the men on the island, but he had to stay in a separate trailer, off the premises of the resort.

The Play

This play is a solo show, meaning that one actor plays several parts in the story, of different ages and genders. The play is historical fiction: that is, the essential facts are drawn from history, all the characters are real people, but the dialogue and some of the details of the plot are from the playwright's imagination, based on what is known of the events.

Characters in the Play

The characters portrayed by the actor include Maurice Ruddick, his wife Norma Ruddick, and his twelve-year-old daughter, Valerie Ruddick. Another featured character is Jack McNeil, a CBC reporter. There are also six other miners in the play: Doug Jewkes, Frank Hunter, Percy Rector, Garnet Clarke, Currie Smith, and Herb ("Pep") Pepperdine.

Synopsis

Scene One

The play opens on Maurice Ruddick, scared and trapped.

A reporter then tells us it's October 23, 1958 in Springhill, Nova Scotia. An earth tremor has caused a "bump" in the Cumberland No. 2 mine. One hundred and seventy-five men are trapped thousands of feet below the earth's surface.

Scene Two

We meet Maurice, before the explosion, and he tells us a bit about his life in Springhill. He talks about the racism he experiences, but he tries not to let it get to him. He always wanted to be a musician, but because he now has twelve kids, he had to move to Springhill to make some money in the mines.

Valerie, Maurice's child, speaks next, giving some history about her mother and father and how they met. Maurice sings his family a song called "Daddy's Gonna Be Home Soon." Valerie makes her father's lunch for the day, including a honey and banana sandwich.

Scene Three

Next, we find Maurice on his way to work, at the Cumberland No. 2 mine on October 23, 1958, a day like any other in Springhill. Maurice makes up a song.

Scene Four

Maurice gets to work. He describes how the men enter the mine and he tells us a bit about working there. In this scene, we meet three other miners: Doug Jewkes, Frank Hunter, and Percy Rector.

Scene Five

Maurice describes the journey down into the mine with his six co-workers. He sings to make the time pass. He describes the environment of the mine. He talks to Percy and offers to share the sandwich made by his daughter Valerie.

Then the first "bump" happens.

Maurice briefly questions his life as a miner and the danger involved. The big bump happens.

Scene Six

The reporter explains what happened at 8:06 pm to the Cumberland Mine No. 2.

Valerie describes what the family was doing when they heard the ominous bump. The whole family is worried about the fate of Maurice.

Maurice comes to after having been knocked out. He sets the scene. Men are screaming. He tries to get everyone alive to come together.

There are six men in the first part of the scene, and then they find Percy, whose arm has been trapped on jagged rocks in the cave ceiling. Now the seven men have to think about how to survive until rescue comes.

Scene Seven

The reporter tells the audience about the men sent to rescue the miners.

The men have been trapped in the mine for three days. They are hungry and their battery packs are dying. They search the pockets of the dead men around them. Frank finds a chocolate bar that everyone shares. Maurice makes sure that Percy eats as well, even though the other men believe he won't make it.

Scene Eight

It's Sunday. The men have begun to despair. Maurice sings a hymn and bangs on the pipes in the mine in the hopes that someone will hear and come find them.

The reporter tells the audience that the last of the men are about to be saved. Valerie sings a song, begging her father to come home.

Scene Nine

We discover that the seven men are still in the mine, even though the people above ground have presumed they are dead. It's Garnet's birthday.

Maurice offers up the last of the food: the sandwich Valerie made for him on the day of the bump.

Percy passes away, and the men become very solemn. Soon afterward, the rest of the men are forced to drink their own urine to stay alive. This drives some of the men over the edge. They are slowly going crazy.

Scene Ten

Halloween night. The reporter interviews Valerie. She still believes, against all odds, that her father is going to come home.

Scene Eleven

Maurice tries to keep his hopes up by singing. Suddenly, the trapped miners hear the tapping of distant rescuers. At last, the men are rescued. Maurice explains what happens to the men once they are free.

Questions and Activities

1. Hold a class discussion about this story as a piece of Canadian history. Why is it important that we tell these stories and remember them? Some topics to bring up: workers' rights to a safe workplace, unsung Canadian heroes, learning the importance of community.

2. Pretend that the year is 1958. Divide your class into groups and give each group a topic. (Some sample topics are: Canadian Politics, Canada-US relations, Economics, Music, Fashion, Schools. Ask students to do some internet/library-based research about Canada at this time; have each group present the information they found.

3. Have your class create collages with images and events that happened that year and post them in class as a reference point for the historical surroundings of the play.

4. Discuss the importance of music and song throughout the play.

   a) When do we sing in our lives? What are some of our favourite songs? How do they make us feel?

   b) Write a song or a poem about Maurice's story.

5. Voice projection is an important part of presenting a play. Beau Dixon, the playwright and original star of *Beneath Springhill*, suggests this exercise to actors who want to make their voices louder: "Put your back against a wall, bend your knees and pretend you are sitting in an imaginary chair. Focus on speaking from your lower back muscles, not your throat. Say words that have vowels and begin with consonants, such as 'baby,' 'bow,' 'boo,' or shout out 'go team go!' in your loudest outside voice. Each sound should start out high and drop down lower, all sounds coming from the lower back, and using the leg muscles to project your voice. This exercise will help make your voice stronger without damaging it."

Try out this exercise and other vocal exercises such as tongue-twisters.

6. What are some ways in which Maurice Ruddick encounters racism in the play?

   a) Do you think there is racism in Canada today? Stage a debate in your class, with each side giving examples to support their position.

   b) Have you personally experienced discrimination because of your race, age, sex, sexual orientation, religion, or for any other reason? If so, write about what happened and how you felt at the time. If you haven't experienced discrimination, find someone who has and interview him or her about what happened and how s/he felt.

7. Beau Dixon writes in the play "We are all the same colour underground." Discuss the different meanings of this statement. The miners were filthy, covered in coal dust and mud. Maurice was already dark and treated differently above ground, but underground the men worked side by side.

8. In December 1995, the Parliament of Canada declared February as Black History Month, following a motion by the Honourable Jean Augustine, M.P., the first Black Canadian woman elected to Parliament. Find out more about Jean Augustine, as well as other Black Canadians who have made contributions to Canadian society and present your findings to the class.

9. Maurice and the other miners use a lot of words to describe what it feels like to be trapped. They also long for freedom.

   a) Create a list of adjectives describing what it must have been like for the men stuck underground. Now create a list of antonyms.

   b) Maurice also uses a few metaphors to describe his situation, create some of your own to illustrate captivity and to illustrate freedom.

10. Cumberland Mine became an official incorporated mine on April 18, 1870, almost 100 years before this disaster.

    a) Get your class to research the mine to find out how many explosions / disasters there were. How many lives were lost to the Cumberland mine?

    b) How many other major mining disasters have happened in Nova Scotia up to this point?

11. Imagine someone you love is stuck in a difficult situation: write a letter to that person using words that will help them feel encouraged, just like Maurice helped his friends while they were stuck in the mine.

12. Ask your class to find out how tall the CN Tower is. How deep was the Cumberland mine where the men were trapped?

13. Maurice Ruddick was awarded "Citizen of the Year" after people heard about his bravery in the mine.

    a) Do you think Maurice was a hero? Discuss why it is important to recognize good deeds and good behaviour.

    b) What are some ways we can all be brave and help people in our community?

    c) Think of something you have done, or that a classmate has done that makes them stand out to you. It could be something small, like lending you a pencil or sharing their food. Make that person an award, to let him or her know that you appreciated it!

14. Maurice's daughter, Valerie, talks about watching television the night that the explosion happened. Here are some of the television shows that she mentions: *I Love Lucy, The Howdy Doody Show, Dragnet* and *Don Messer's Jubilee.*

    Ask your class to look up clips of these shows on the Internet.

a) Have a conversation comparing entertainment for kids in 1958 with entertainment for kids now.

b) Divide your class into groups to write a radio or television news report detailing the events of the Springhill mining disaster. They can have fun making up their own commercials for products sold in the '50s in Canada to air during the news report.

Glossary

Bump:
A violent dislocation of the mine workings, which is attributed to severe stresses in the rock surrounding the workings.

Coal:
A black or dark-brown combustible mineral substance consisting of carbonized vegetable matter, used as a fuel.

Draglines:
An excavation system used to remove layers of rock and soil covering a coal seam. The dragline casts a wire rope-hung bucket a considerable distance, collects the dug material by pulling the bucket toward itself on the ground with a second wire rope (or chain), elevates the bucket, and dumps the material on a spoil bank, in a hopper, or on a pile.

Hydrogen sulphide:
Also called: sulphuretted hydrogen. A colourless, poisonous soluble, flammable gas with an odour of rotten eggs.

***I Love Lucy, The Howdy Doody Show, Dragnet*:**
Popular television shows from the 1950s.

Jacks and shears:
Tools used for mining

Kerosene lamp:
(Also known as a paraffin lamp) A type of lighting device that uses kerosene as a fuel. Kerosene lamps have a wick as a light source, protected by a glass chimney or globe; lamps may be used on a table, or hand-held lanterns may be used for portable lighting.

Mulatto:
A person of mixed Caucasian and African ancestry.

Lamp room:

A room or building at the surface of a mine for charging, servicing, and issuing all cap, hand, and flame safety lamps. Also known as lamp cabin; lamp station.

Pithead:

Entrance to the actual mine

Pomade:

A greasy and waxy substance that is used to style hair. Pomade makes hair look slick, neat and shiny.

Timber road:

Large, unpaved roads used to haul timber to the mine. Timber was used to form support beams around the miners as they were digging.

Wash house:

The miners' wash and change area.

Resources

Brown, Roger David. *Blood on the Coal: The Story of the Springhill Mining Disaster* (Nimbus Publishing, 1976).

CBC News. "Springhill Mining Disasters." https://www.cbc.ca/archives/topic/springhill-mining-disasters.

Greene, Melissa Fay. *Last Man Out: The Story of the Springhill Mine Disaster* (Harcourt, Inc. 2003).

McKay, Cheryl. *Spirit of Springhill: Miners, Wives, Widows, Rescuers & Their Children Tell True Stories of Springhill's Coal Mining Disasters* (Purple Penworks, 2014).

McKay, Ian. "The Realm of Uncertainty: The experience of work in the Cumberland Coal Mines 1873-1927" *Acadiensis*, 16 no.1 (Autumn 1986): 3–7.

Stanbridge, Joanne. *Maurice Ruddick: Springhill Mine Survivor* (Pearson Education Canada 2005).

www.gov.ns.ca/nsarm/virtual/meninmines